HOME ALONE – AND HAPPY!

Essential life skills for preventing separation anxiety in dogs and puppies

Kate Mallatratt

Hubble & Hattie

Hubble & Hattie

The Hubble & Hattie imprint was launched in 2009 and is named in memory of two very special Westie sisters owned by Veloce's proprietors.

Since the first book, many more have been added to the list, all with the same underlying objective: to be of real benefit to the species they cover, at the same time promoting compassion, understanding and respect between all animals (including human ones!)

Hubble & Hattie is the home of a range of books that cover all-things animal, produced to the same high quality of content and presentation as our motoring books, and offering the same great value for money.

More great Hubble & Hattie books!
Among the Wolves: Memoirs of a wolf handler (Shelbourne)
Animal Grief: How animals mourn (Alderton)
Babies, kids and dogs – creating a safe and harmonious relationship (Fallon & Davenport)
Because this is our home ... the story of a cat's progress (Bowes)
Bonds – Capturing the special relationship that dogs share with their people (Cukuraite & Pais)
Camper vans, ex-pats & Spanish Hounds: from road trip to rescue – the strays of Spain (Coates & Morris)
Canine aggression – how kindness and compassion saved Calgacus (McLennan)
Cat and Dog Health, The Complete Book of (Hansen)
Cat Speak: recognising & understanding behaviour (Rauth-Widmann)
Charlie – The dog who came in from the wild (Tenzin-Dolma)
Clever dog! Life lessons from the world's most successful animal (O'Meara)
Complete Dog Massage Manual, The – Gentle Dog Care (Robertson)
Confessions of a veterinary nurse: paws, claws and puppy dog tails (Ison)
Detector Dog – A Talking Dogs Scentwork Manual (Mackinnon)
Dieting with my dog: one busy life, two full figures ... and unconditional love (Frezon)
Dinner with Rover: delicious, nutritious meals for you and your dog to share (Paton-Ayre)
Dog Cookies: healthy, allergen-free treat recipes for your dog (Schöps)
Dog-friendly gardening: creating a safe haven for you and your dog (Bush)
Dog Games – stimulating play to entertain your dog and you (Blenski)
Dog Relax – relaxed dogs, relaxed owners (Pilguj)
Dog Speak: recognising & understanding behaviour (Blenski)
Dogs just wanna have Fun! Picture this: dogs at play (Murphy)
Dogs on Wheels: travelling with your canine companion (Mort)
Emergency First Aid for dogs: at home and away Revised Edition (Bucksch)
Exercising your puppy: a gentle & natural approach – Gentle Dog Care (Robertson & Pope)

For the love of Scout: promises to a small dog (Ison)
Fun and Games for Cats (Seidl)
Gods, ghosts, and black dogs – the fascinating folklore and mythology of dogs (Coren)
Helping minds meet – skills for a better life with your dog (Zulch & Mills)
Home alone – and happy! Essential life skills for preventing separation anxiety in dogs and puppies (Mallatratt)
Know Your Dog – The guide to a beautiful relationship (Birmelin)
Letting in the dog: opening hearts and minds to a deeper understanding (Blocker)
Life skills for puppies – laying the foundation for a loving, lasting relationship (Zuch & Mills)
Lily: One in a million! A miracle of survival (Hamilton)
Living with an Older Dog – Gentle Dog Care (Alderton & Hall)
Miaow! Cats really are nicer than people! (Moore)
Mike&Scrabble – A guide to training your new Human (Dicks & Scrabble)
Mike&Scrabble Too – Further tips on training your Human (Dicks & Scrabble)
My cat has arthritis – but lives life to the full! (Carrick)
My dog has arthritis – but lives life to the full! (Carrick)
My dog has cruciate ligament injury – but lives life to the full! (Häusler & Friedrich)
My dog has epilepsy – but lives life to the full! (Carrick)
My dog has hip dysplasia – but lives life to the full! (Häusler & Friedrich)
My dog is blind – but lives life to the full! (Horsky)
My dog is deaf – but lives life to the full! (Willms)
My Dog, my Friend: heart-warming tales of canine companionship from celebrities and other extraordinary people (Gordon)
Office Dogs: The Manual (Rousseau)
One Minute Cat Manager: sixty seconds to feline Shangri-la (Young)
Ollie and Nina and ... daft doggy doings! (Sullivan)
No walks? No worries! Maintaining wellbeing in dogs on restricted exercise (Ryan & Zulch)
Partners – Everyday working dogs being heroes every day (Walton)

Puppy called Wolfie – a passion for free will teaching (Gregory)
Smellorama – nose games for dogs (Theby)
Supposedly enlightened person's guide to raising a dog (Young & Tenzin-Dolma)
Swim to recovery: canine hydrotherapy healing – Gentle Dog Care (Wong)
Tale of two horses – a passion for free will teaching (Gregory)
Tara – the terrier who sailed around the world (Forrester)
Truth about Wolves and Dogs, The: dispelling the myths of dog training (Shelbourne)
Unleashing the healing power of animals: True stories about therapy animals – and what they do for us (Preece-Kelly)
Waggy Tails & Wheelchairs (Epp)
Walking the dog: motorway walks for drivers & dogs revised edition (Rees)
When man meets dog – what a difference a dog makes (Blazina)
Wildlife photography: saving my life one frtame at a time (Williams)
Winston ... the dog who changed my life (Klute)
Wonderful walks from dog-friendly campsites throughout the UK (Chelmicka)
Worzel Wooface: For the love of Worzel (Pickles)
Worzel Wooface: The quite very actual adventures of (Pickles)
Worzel Wooface: The quite very actual Terribibble Twos (Pickles)
Worzel Wooface: Three quite very actual cheers for (Pickles)
You and Your Border Terrier – The Essential Guide (Alderton)
You and Your Cockapoo – The Essential Guide (Alderton)
Your dog and you – understanding the canine psyche (Garratt)

Hubble & Hattie Kids!
Fierce Grey Mouse (Bourgonje)
Indigo Warrios: The Adventure Begins! (Moore)
Lucky, Lucky Leaf, The: A Horace & Nim story (Bourgonje & Hoskins)
Little house that didn't have a home, The (Sullivan & Burke)
Lily and the Little Lost Doggie, The Adventures of (Hamilton)
Wandering Wildebeest, The (Coleman & Slater)
Worzel goes for a walk! Will you come too? (Pickles & Bourgonje)
Worzel says hello! Will you be my friend? (Pickles & Bourgonje)

www.hubbleandhattie.com

Disclaimer
Please always follow the manufacturer's guidelines with regard to providing toys and feeding your dog. If in doubt please speak to your own veterinary surgeon. Remember that no toy is indestructible: please ensure your dog is safe when you leave him with any of the *Home alone – and happy!* tools discussed. Raw bones, in particular, are high value items to many dogs: please read the advice about guarding before you give your dog a raw bone.

Having put into practice the advice in this book, if you are at all worried about your dog, or if your dog is still displaying signs of anxiety when left alone, it is recommended you seek the advice of a professional behaviourist on veterinary referral.

First published May 2016 by Veloce Publishing Limited, Veloce House, Parkway Farm Business Park, Middle Farm Way, Poundbury, Dorchester, Dorset, DT1 3AR, England. Reprinted February 2017 and October 2017. This edition published June 2019; reprinted September 2019. Fax 01305 250479/email info@hubbleandhattie.com/web www.hubbleandhattie.com ISBN: 978-1-787115-59-0 UPC: 6-36847-01559-6 © Kate Mallatratt & Veloce Publishing Ltd 2016, 2017 & 2019. All rights reserved. With the exception of quoting brief passages for the purpose of review, no part of this publication may be recorded, reproduced or transmitted by any means, including photocopying, without the written permission of Veloce Publishing Ltd. Throughout this book logos, model names and designations, etc, have been used for the purposes of identification, illustration and decoration. Such names are the property of the trademark holder as this is not an official publication. Readers with ideas for books about animals, or animal-related topics, are invited to write to the editorial director of Veloce Publishing at the above address. British Library Cataloguing in Publication Data – A catalogue record for this book is available from the British Library. Typesetting, design and page make-up all by Veloce Publishing Ltd on Apple Mac. Printed in India by Replika Press.

Contents

Acknowledgements and Dedication 4
Foreword by Nando Brown 5
Introduction 7

1: What is separation anxiety? 11
   Signs to look out for 11
   Why prevention is better than cure 13
   Why punishment should never be used 21

2: Managing emotions 22
   Exploring emotions 22
   Analysing emotions, and how these may
     outwardly display 32
   Managing emotions on departure 32
   Establishing solid foundations 38
   Building anticipation 38

3: Building independence 40
   Mealtime training: feed him and leave him! . . . 40
   Eating for pleasure 43
   Food ingredients 48
   Fussy eaters 50
   Feeding raw bones and bone safety 50
   Teaching 'settle on bed' 52
   Building independence 52

4: Managing the environment 59
   Beds and sleep quality 59

Room temperature 62
Reducing arousal 63
Movement restriction 64
The crate: your dog's bedroom 64
The sound of music 66
Scent 69

5: Home alone and happy toolbox 72
   Environmental cues 74
   Key training reminder 74
   Kongs™ and other interactive toys 74
   Toys 75
   Should we feed our dog every time we
     leave him? 75
   Signs that our best friend is happy 83

6: Assessment and worksheets 84
   Comfort evaluation worksheets 84

Appendix: Resources and further reading; references . . 87
Rogues' Gallery: the Cast 89
Index 96

Acknowledgements and Dedication

My sincere thanks go to my family, friends and colleagues who have made this book possible, especially Sherri Steel, and my husband, Paul Mallatratt. Very special thanks are extended to Cheryl Murphy for capturing such beautiful and expressive photographs of the many dogs included. My thanks also go to everyone who has allowed us to photograph their dog: your generosity has made this book possible.

Dedication

This book is dedicated to all those who have influenced my thinking over the years, helped my learning evolve, and believed in me.

Above all, my gratitude goes to the many dogs who have touched my life and taught me so much: dogs who have shown me that, if you have the ears to listen, they will talk.

He is your friend, your partner, your defender, your dog. You are his life, his love, his leader. He will be yours, faithful and true, to the last beat of his heart. You owe it to him to be worthy of such devotion.

Anon

Foreword by Nando Brown

Separation anxiety is my personal nemesis. I hate it! I dislike it because it's such a hard problem to deal with: it's hard for me as a trainer, it's hard for the owner, but most of all it's so very hard for the dog.

It's hard for me as a trainer, because once your dog has a deep-rooted anxiety issue, the amount of time and complexity of work that needs to be invested is enormous. I need to explain the science behind it in terms that can be easily understood and applied practically. It's hard for the owner, because sometimes it can feel overwhelming. The frustration of repeatedly coming home to destroyed possessions, including furniture, walls and doors, without knowing why or what you can do to prevent it, can be too much for some. And it's hard for the dog because that emotion of utter fear, anxiety or desperation is a very real one.

If you suffer from arachnophobia then you will understand exactly how debilitating fear can be when you spot a spider heading in your direction. Imagine, if you will, that your best friend controlled the spider, and they gave you a number of indications that they were about to put you in a situation where you had no control over the spider's movements. The fear would begin when you thought about being left with the spider – well before you saw the spider! But no matter what you said, or how anxiously you behaved, you simply could not communicate to your friend that you didn't want the spider near you. Yet, day after day, they left you with the spider and no escape route! The immense fear you would feel in this situation is exactly the same as the fear that a dog feels when his anxiety levels rise out of control; often beginning when we get ready to leave the house.

The idea behind *Home alone – and happy! – prevention –* is one that is very important to me. Building a

Tye Prints Photography www.tyeprintsphotography.com

Home alone – and happy!

dog's self-confidence sufficiently to allow him to feel secure in his own company is vital. This book will give you the tools to make your dog a happy, healthy and confident home-alone canine; one who can be left without fear of the 'spider' lurking, allowing you some control over your free time once more!

Congratulations to you! Congratulations because, instead of turning to the local pub 'expert,' your neighbour or social media for advice, you have just begun a journey with a book that has been written by a certified, well-respected professional in canine behaviour management. Kate is a founder member of the International Canine Behaviourists,

and she will help you learn to use the tools that will enable you to have a happy, home-alone hound!

Separation anxiety is distressing for both the dog and his owner. The focus of Kate's book is preventative 'medicine,' and it is full of practical tips. I highly recommend it for any dog owner, but especially the new puppy owner who would like to be able to leave their dog home alone – and happy!

Nando is a behaviour expert and instructor for the Institute of Modern Dog Trainers. He co-founded In the Dog House training school in Spain in 2009, where he was Head Trainer for seven years, prior to moving back to the UK. He is a sought-after international speaker, regularly appearing at conferences and on television and radio, discussing dog behaviour issues. He runs the very popular Facebook page Incredimal, which follows the training of his Belgian Malinois, Fizz, from puppyhood.

INSTITUTE OF MODERN DOG TRAINERS

Introduction

Owning a dog is one of life's great pleasures: dogs enrich our lives immensely, and are our best friends. They are sentient beings with thoughts, feelings and emotions, likes and dislikes, and it is for these very reasons that their behaviour can, however, occasionally go awry. Managing certain behaviour problems can be challenging, infringe enormously on daily life, and be time-consuming, stressful and even costly, especially if professional help is sought. Separation anxiety is one such behaviour problem.

A dog with separation anxiety (one who is unable to cope with being alone) can severely affect both your own and your dog's quality of life. Your dog may become distressed as a result (and may harm himself in his attempts to escape his solitary confinement), and you may receive complaints from neighbours if he is vocalising his discomfort, as well as possible damage to your property. Continuing stress is not good for your dog's long-term health, either, and may result in chronic or acute medical conditions. Taking a little time to teach your dog or puppy how to enjoy his own company before problems arise will enable you both to enjoy some quality time when apart.

People generally leave it too long before seeking help in this situation. Many trainers and behaviourists receive desperate telephone calls when an owner is at the end of their tether, by which time the problem has often been present for months, if not years, gradually worsening in that time. The dog may have been suffering for months before his owner becomes aware of the problem because it has escalated to physical injury to the dog, nuisance barking, or damage to property.

In puppy classes, people proudly tell trainers that their dog can sit. What we would really like to hear is how proud they are that their puppy jumps around with joy in

Opening our home – and our heart – to a dog enriches our lives immeasurably.

anticipation of a stuffed Kong™ when they are leaving the house! Most dogs can easily and quickly be taught to sit, but a one-year-old adolescent with an eight month history of worsening separation anxiety will require a great deal of behaviour modification work to resolve the problem.

It's important to realise that an obedient dog isn't necessarily an emotionally stable one. In conjunction with basic obedience training and socialisation, many owners should place more emphasis on teaching their dog

Home alone – and happy!

Sit is one of the first behaviours new owners teach, practicing it many times a day. Home-alone practice also needs to happen every day to teach a puppy the skills he needs to enable him to enjoy his own company.

Puppyhood is a perfect time to focus on modifying a dog's emotions to happy ones when he is left alone, teaching him coping skills before it becomes a problem. Relaxation skills and emotional contentment – for those times when your dog is home alone – will help: don't assume he has or will develop these skills of his own accord. Doing nothing and leaving it to chance may lead to your dog experiencing growing disappointment when you leave, snowballing to anxiety in future. Your leaving the house should be a positive experience for your dog – and not a negative one, at the very least. Don't wait until your dog runs into trouble before teaching him the skills he needs: get it right from day one!

This book takes you through the necessary steps to help prevent separation anxiety, especially in puppies, before it starts, and provides the tools for teaching both dogs and puppies to enjoy their own company before any underlying anxiety has a chance to escalate. This approach can prevent much anguish and frustration for you and your dog, and time and money for you, should a problem develop. Avoidance of a behaviour problem should be a primary motivation for teaching *Home alone – and happy!* skills, although simply wishing to improve your dog's quality of life when he is home alone is also reason enough.

The techniques in this book are intended primarily as prophylactic aids for dogs and puppies who do not suffer from separation anxiety, or those who experience very mild separation anxiety/over-attachment. In this instance, your dog may be disappointed when you leave, but, if left with a tasty morsel, will eat it before your return. A dog with a moderately severe case will probably be too anxious to eat (see 'signs to look out for' on page 11 for more information), although some anxious dogs will wolf down food.

Even if you own an adult dog or have recently brought a rescue dog into your home who doesn't appear to have any problems with being left alone, why not take steps to secure your dog's future emotional comfort?

A dog's underlying anxiety about being left alone may seep into other areas of his life, making him generally more sensitive and reactive. Emotional states can fluctuate – aren't we all a little touchy when we've had a bad day at work? A home-alone dog who has no coping skills has a bad day at work *every* day. After suffering a traumatic life event, a person can become more vulnerable and sensitive

important life skills with an emotional element in those early months, such as being content in their own company. A lack of skills such as this can result in anxiety, and severely impact on a dog's quality of life – and be stressful for his owner.

Dogs and puppies don't often 'grow out of' separation anxiety, and, instead, usually become gradually more anxious. Taking your dog for a long walk before you leave him is not generally a solution for relieving separation anxiety; nor is hoping he will eventually get used to being alone.

to stress, and certain situations – or even just the thought of them – can trigger anxiety and panic attacks, and even post traumatic stress disorder. Similarly, a dog with developing separation anxiety will be anxious before we even leave the house, and even after behavioural modification work, will have an underlying vulnerability that may resurface in times of stress.

A dog with moderate to severe separation anxiety should be seen by a professional behaviourist, and, if in any doubt at all about your dog's health or behaviour, seek a veterinary assessment in order to rule out any underlying medical cause. Be aware that the wrong advice can make a behaviour problem worse. If you are at all concerned about your dog's behaviour, don't wait until the problem escalates: find a suitably qualified behaviourist, who should –

• be fully qualified and insured
• be a member of a professional body
• use positive reinforcement training
• adhere to a strict code of conduct
• provide evidence of continuing professional development

– to properly diagnose the problem and come up with a tailored behaviour modification plan. Walk away from any trainer who tells you your dog is being 'dominant' or that you need to be the 'pack leader,' as this is outdated science that appears to be flawed.

What is _Home alone – and happy!_ training, and how will it help my dog?
Simply put, it's a toolbox of skills which are designed to teach and maintain relaxation in your dog or puppy to enable him to occupy the time you are away, pleasantly and without fear. It includes lifestyle choices for you and your dog which, if taught correctly, enable your dog's time alone to become a positive emotional experience and not a negative one. _Home alone – and happy!_ training helps prevent emotional trauma by laying strong foundations for an emotionally well-balanced, healthy and happy dog for life, allowing you to feel confident about leaving him, secure in the knowledge that he is able to enjoy his own company.

How to use this book
Aimed at dogs and puppies who are not experiencing separation anxiety problems, this book will advise how to enrich your dog's environment when he is on his own at home.

You will learn what separation anxiety is, how to assess what your dog may be feeling about being left, and understand why teaching puppies to enjoy their own

Anxiety about being left often has its roots in puppyhood, when a pup's first experiences of being left alone are frightening.

Home alone – and happy!

company at the outset is important. Answering questions about your dog's level of comfort before training, and then assessing him again afterward, will provide valuable knowledge about how emotionally comfortable your dog is when alone.

- Chapter 1 explains how a diagnosis of separation anxiety is made, what signs to look for, and why punishment should never be used

- Chapter 2 considers the relationship we have with our dog, and looks at ways we can begin to teach him to enjoy his own company. It discusses how to pair rewards with departure without negatively sensitising our dog to the reward

- Chapter 3 describes how to build independence in our dog, choose good quality foods, and feed him in ways that

increase his eating pleasure, and the time he spends doing so. The problem of fussy eaters is discussed

- Chapter 4 covers environmental influences on relaxation

- Chapter 5 looks at the *Home alone – and happy!* toolbox: the tools available to help our dog relax in his own company and positively come to terms with being on his own at home

- Comfort evaluation sheets in Chapter 6 ask key questions to help assess your dog's level of contentment before and after training

Please note that, for ease of reading, the text refers to dogs as male throughout: however, female is implied at all times. The term 'dog' refers to an adult, puppy, rehomed/rescue/breed rescue, mixed breed or pedigree dog.

Separation anxiety is a complex disorder, which, once established, can require professional help to resolve.

1: What is separation anxiety?

Separation anxiety is the inability of a dog to cope emotionally when apart from his owner, evidenced by varying levels of discontent, which often give rise to unwanted behaviour. According to one UK study, the number of dogs suffering from separation anxiety could be as high as around 29 percent,[1] and many more dogs suffer, but do not get the help they so desperately need.

Severe separation anxiety is a serious condition, which can cause dogs to follow their owners around the house, as they are simply unable to cope with being alone. Other signs may include barking, destructiveness, anxiety (drooling, panting, whining, pacing, trembling, shaking), house soiling, inappetence and attempts to escape. Some dogs display few outward signs of distress, but may become withdrawn and depressed when alone. Many owners do not realise that their dog isn't coping well alone until a neighbour complains about the barking, or they return to find their dog has been destructive, and/or soiled the house. None of these dogs is being wilfully bad or spiteful because he has been left: it's that he is distraught without his owner, and does not know how to cope with this feeling.

Signs to look out for

What are the signs that let us know our dog might be anxious when alone?

More often than not, a dog with the beginnings of separation anxiety may follow us upstairs to the bathroom, and lie outside the door, possibly scratching to come in. He will often follow us around the house, and may look worried and uncomfortable when we leave, perhaps letting out a few whimpers or barks. He may run to the window and watch us drive away. He may not eat food that has been left for him, since anxiety causes appetite suppression. Our dog

'Waiting for mum.' Many dogs will watch through the window for our return, too anxious to eat, relax or sleep. Dogs who are lonely and worried don't always display outward signs of this: they simply suffer in silence.

isn't annoyed that we have gone; he is worried about being on his own.

This anxiety could kick in when we prepare to leave the house, or even when we get dressed for work, pick up the car keys, or turn on the radio 'for company.' All of these actions can become learned predictors to our dog, letting him know that we are leaving, causing anxiety to build.

On our return, a dog or puppy who is worried about being left alone is often frantic with relief. He may be in the same place we left him – by the door or window – and his frenetic welcome should not be mistaken for pleasure

1 Bradshaw J W S, McPherson J A, Casey R A, Larter I S. Aetiology of separation-related behaviour problems in domestic dogs. *Veterinary Record*, 151, 2002, pp43-46.

Home alone – and happy!

Our dogs love spending time in our company and being near us.

that we are back. In this instance, his greeting is too intense and goes on for too long, and he takes a long time to settle again, possibly not letting us out of his sight. These are all indications that our dog is finding it hard to cope alone.

We all want our dogs to spend time with us and be near us. When we settle down in front of the television in the evening, say, sit down with a good book, or even work at the computer, our dogs often settle by our side. This is normal behaviour, but do not mistake over-attachment for affection: a puppy who follows his owner from room to room as he is unable to happily occupy himself is not developing a healthy relationship with his owner. Owners who are retired or who work from home may be at particular risk of having their puppy become over-attached, since they spend a lot of their day in each other's company. While there is nothing intrinsically wrong with spending much of the day with our dog, we need to be aware of the possible consequences:

Being out and about with our dogs is a favourite pastime for them, and builds a unique connection.

our dog may not develop the ability to cope alone if never left, and separation anxiety may result. It is better to teach a puppy how to enjoy his own company right from the start.

Some dogs do not leave evidence of separation anxiety. Research led by Dr Emily Blackwell at Bristol University School of Veterinary Sciences, found that although dog owners claimed their animals did not suffer from

Disappointment that we are leaving the house without him can grow into anxiety if our dog's emotional needs go unfulfilled when he is alone.

separation stress, video and salivary cortisol measurements told a different story: over 80% of dogs showed either behavioural or physiological signs of stress when left alone.[2]

A dog who has a healthy and balanced relationship with his owner will enjoy the time they spend together very much, but will be comfortable when apart, too. When his owner gets up to leave the room, he will continue to doze, perhaps watching out of one eye; aware that his owner has left but secure enough not to worry, or feel the need to follow (unless he goes into the kitchen, of course!). While he may prefer to curl up at his owner's feet, he can be content to lie further away also, and does not feel the need to follow his owner from room to room, in constant contact: he is secure and independent.

When his owner dons coat and shoes to leave the

house, a well-balanced dog may wander off to his bed to sleep, or happily wait in the kitchen for his treat. On our return, he will be delighted to see us but not frantic; checking out any shopping bags we carry in, and perhaps bringing us a toy. These are all normal signs that our dog is pleased we are home, and he is able to settle down again quickly.

Why prevention is better than cure
Prevention is better than cure because first learning sticks, and behaviours that we don't want to encourage – such as barking or chewing – become ingrained if practised regularly. First learned behaviours are often those which are reverted to in times of stress, and, regardless of whether or not these are desirable, repetition means that muscle memory is built.

Muscle memory growth takes place not, as the name might suggest, in the muscles but in the brain. When dogs do not have the necessary neural connections in the brain

A dog who can enjoy his own company will be content to stay on his own – although he may still choose to lie somewhere where our scent lingers, as this provides a source of comfort.

2 Dr Emily Blackwell (2014) *pers comms* 8 March 2016. Research unpublished. Bristol University School of Veterinary Sciences.

13

Home alone – and happy!

for a new activity, learning is hard work, and the activity concerned requires focus, co-ordination and concentration, and at first appears awkward. The more the activity is repeated, however, the more the brain becomes 'hard-wired' to perform the activity smoothly. Neurotransmitter chemicals stimulate brain cells related to the activity to grow dendrites (a short branched extension of a nerve cell), and eventually these form a neural pathway. After many, many repetitions, the activity becomes second nature.

This is an important consideration because, if we let our dogs practise unwanted behaviour, neural pathways (muscle memory) are built in the same way as if they are practising desirable behaviour. Changing behaviour that has been practised, often for many months or even years, before we invest time in teaching an alternative, more acceptable, behaviour, is much more difficult. Old habits die hard ...

The following example illustrates just how hard it is to change an ingrained practice that we are probably all familiar with: riding a bike. An engineer tried to ride a bicycle that had the handlebars swapped around. When he steered to the right, the bike's wheel went to the left, and when the handlebars were turned to the right, the wheel steered to the left. Having ridden a bicycle for some twenty-five years, it took the engineer eight months of practising for five minutes a day to re-learn how to ride, and even then, without good focus and concentration, he reverted to previous learning.

A dog whose anxious behaviour has gradually increased over months and possibly years will find it much more difficult to learn new behaviour, and a new, positive, conditioned emotional response associated with that new behaviour, than a dog who has learned the correct behaviour from puppyhood, or who has not yet developed anxiety about being left alone. It's important to note that chewing, in the absence of any other sign of separation anxiety, is not necessarily a symptom of separation anxiety; it can be due to boredom, or simply that something appealing was available!

Any breed or mixed breed of dog can suffer from separation anxiety, and genetics heavily influence behaviour. Some dogs are more sensitive than others, and more easily form unhealthy over-attachments to their owners.

Separation anxiety may also peak when a dog is already anxious about other stressors, such as a house move,

Separation anxiety doesn't usually appear out of the blue, and often grows gradually, starting in puppyhood and worsening until outward signs become apparent.

traumatic life event, loss of an owner, or a child leaving home.

Although these stressors may be a trigger, the anxiety is unlikely to occur out of the blue: it's more inclined to have been present in a mild form prior to the trigger.

Behaviour problems can worsen over time, and it's often the case that an owner seeks advice only when the problem has become serious and inconvenient, rather than recognising the first signs and taking action then. Preventing separation anxiety in the first place is by far the best approach! Puppy owners sometimes wonder why they are teaching *Home alone – and happy!* skills when their puppies appear perfectly fine, but this is because it's easier to teach

Any breed, or crossbreed of dog, can suffer from separation anxiety, although some can be more susceptible than others. Even in a multi-dog household, a dog may suffer from separation anxiety if he is over-attached to his owner.

Home alone – and happy!

a puppy to enjoy his own company *before* he can become anxious about being left alone.

Even if your puppy isn't vocalising, being destructive or soiling in the house, he may still be disappointed when you leave, and disappointment can snowball into anxiety, then develop into separation anxiety. The subtle first signs of separation anxiety may well be missed, or may not start or develop until a puppy reaches adolescence: a time when he becomes more aware of his environment – and of being left alone. This is also when a lot of changes occur in the brain as he matures, and dogs often display a second fear period around this time, when they become more sensitive to their surroundings.

Separation anxiety can creep up on the unsuspecting owner, who might be of the opinion that their puppy is fine when left alone. Closer examination often reveals early indications of potential problems. For example –

- A puppy may follow us from room to room
- He may appear to be fast asleep but, as soon as we move, he follows
- He may cry if a door accidentally closes in his face, or whine if shut in a crate
- He may sleep on our feet, and seek to be physically close at all times, taking umbrage at enforced separation
- He may be unable to sleep alone at night-time, or is spending the night on our bed. (Sleeping on our bed (unless he has resource-guarding issues), or in our bedroom, isn't necessarily a problem, although, in a dog with over-attachment issues, this may create further dependency.)

Unless your dog has resource guarding or over-attachment issues, allowing him to sleep on your bed should be fine, and some dogs enjoy the comfort of their owner's bed when home alone, with its reassuring scent. However, be aware that beds are often important areas of the house for many dogs, and may be guarded.

Some dogs show no outward signs of anxiety on our departure, especially if there is another dog in the house, providing an emotional 'anchor' and masking the anxiety. A dog may, however, still suffer from separation anxiety in a multi-dog household if the cause of it is owner over-attachment, so getting another dog for company isn't necessarily a solution.

If the reason for the separation anxiety is owner over-attachment, acquiring a second dog may not achieve

continued page 20

Compatible companions form strong bonds and enjoy mutual interests ...

... and often mirror each other's body language, and happily share toys.

How will they cope if, never having been apart from each other, one of your dogs has to stay at the vet's for a day ...?

Home alone – and happy!

Dogs who have previously been perfectly fine when home alone may need more support as they approach their twilight years.

a satisfactory outcome. For a start, the dogs have to get on well and become compatible companions, which must happen naturally and without pressure: after all we wouldn't like to share our house with just anyone! An older dog may not necessarily appreciate having a young puppy in the house, especially if the anxiety about being left alone is due to cognitive dysfunction.

Some dogs can become great companions, of course, but think carefully before you get another dog, and consider the needs, breed and age of your current dog. You will often see two gun dogs, for instance, sharing the same sniffs on a walk, so do think carefully if looking for a second companion. Compatible companions will often mirror each other's behaviour.

It is good practice to teach individual dogs in a household the skills they need to be alone, and get them used to enjoying their own company from an early age. For example, feed separately at mealtimes, gradually extending the time and building the skills they need to be apart, until one dog can be happily left alone with his treat, while a second dog is taken for a walk.

As dogs age, they may need more support when left alone. As previously mentioned, anxiety about being left may be due to cognitive dysfunction, and dogs so affected may seek owner-comfort more often. Their physical needs may change: they may be on pain medication, need to toilet more often, and be more sensitive to temperature. Always discuss any changes in your older dog's behaviour with your veterinary surgeon, as it is important to rule out underlying medical conditions.

Why punishment should never be used

Any attempt to punish your dog is likely to make the anxiety worse, and, in any case, only positive reinforcement should be used. Being alone is extremely stressful for your dog, and a reprimand will only add to his anxiety, because he does not have the mental capability to understand what it is he's done 'wrong.' Dogs simply do whatever make them feel better. Punitive measures using aversive products such as citronella spray collars to prevent barking when alone are not recommended because they are unkind, and do not treat the root cause of the problem, in any case. An anxious dog needs to relieve his anxiety somehow, and if prevented from doing so one way, he may find another outlet that has more serious consequences, such as self-mutilation (chewing his paws), and may even become depressed if unable to resort to his usual coping mechanism.

Likewise, 'time out' in a crate, room or garden is going to make him even more anxious, since being alone – and away from you – is why he is anxious in the first place! He is not being naughty; he simply needs you around to help him to feel secure.

Implementing any kind of 'pack leader' rank reduction programme is unlikely to help your dog's separation anxiety, either, and could even be harmful. Research shows that the dominance theory is seriously flawed: your dog is not suffering from separation anxiety because he wants to be in charge, or dominant; he is simply anxious about being alone, just as a child might be scared of the dark and need comforting. Withdrawing your attention from a dog who is so desperate to see you he has chewed the sofa as a way of relieving his anxiety, is going to make him even more confused and anxious.

Dogs with severe separation anxiety may well require veterinary assessment, and possibly anti-anxiety medication, combined with a tailored behaviour modification programme. Sadly, even then, a cure isn't guaranteed.

2: Managing emotions

Exploring emotions

An emotion could be described as a state of mind that reflects inner feelings such as happiness and sadness. Emotions are often outwardly expressed in our behaviour, and can be triggered internally through thoughts and feelings, and externally through objects, people or animals, and significant events: music is a very powerful medium for stirring emotion, for example.

Dogs have emotions, too, which are, perhaps, more complex than we fully appreciate. Emotions are necessary for survival. They help an animal to avoid aversive situations, and seek out valuable resources. The amygdala, a primitive

A penny for his thoughts? How do you think this dog is feeling ...?

part of the brain, which appears to play an important function in processing emotion, is present in both our brain and that of our dogs.

Behaviour has a direct impact on emotion. For example, when happily chewing a bone, a dog will experience a release of endorphins: feel-good chemicals which produce a feeling of contentment and well-being. Conversely, emotion can influence behaviour: a dog who is unhappy at home alone, will experience a release of stress hormones – adrenaline and cortisol – and may chew furniture, for example, in a bid to alleviate the feelings of anxiety that result.

Although there are breed similarities, every dog is unique, of course, with individual needs, and each will react differently on an emotional level to being left alone at home. Our dogs' coping skills can vary due to genetic make up, physical attributes, any medical conditions, and his history of learned behaviour.

Evaluating emotional states in animals is complicated. Much of the research into canine emotion measures negative emotions such as fear, anxiety and frustration, because these are easier to accurately assess. Animal welfare organisations and pharmaceutical interests tend to focus on this area, and research into positive emotional states is limited. Most studies on human emotion involve language-based tasks, but it is, of course, impossible to ask a dog how he feels. One indicator of a positive state of mind is play, which often occurs when an animal's other basic needs have been satisfactorily met.

One dog may suffer disappointment when home alone, but be able to cope with this, while another may become so fearful that he self-mutilates, vocalises, and damages property. There are a whole range of emotions in-between to consider – relaxation, disappointment and fear, for example – and how a dog feels is fluid, and changes from hour-to-hour; even minute-to-minute. We cannot be certain that the emotions of happiness and joy that we feel – or, indeed, sadness and anxiety – are the same as those

Our dogs express many emotions. The following pictures illustrate some of the positive emotions our canine friends express –

Joyfully chasing a ball ...

... delightedly cavorting in water ...

our dogs feel. Some dogs outwardly grieve at the loss of a companion, whilst others do not appear to give a jot. A dog derives obvious enormous pleasure and information from scenting his environment, and we can only imagine how that must feel. Some of the emotions that a home-alone dog feels may be conjecture on our part, and anthropomorphic (ascribing human characteristics), but, in the absence of easily accessible ways to measure our dog's emotions, we have little choice but to empathise with how he may be feeling, based on knowledge, experience, observation, and comparison with our own emotions.

The catalyst for teaching home-alone skills should be the compassion we feel for our dogs when they are alone, rather than waiting until a problem arises.

We often hear the claim 'He can understand every word we say!' Of course, a dog cannot possibly do this, and to pretend otherwise is not sensible. Embracing, understanding and celebrating the differences between us and our dog will, however, help to overcome many obstacles to communication, and recognising his needs as a unique species will help to provide more satisfactorily for his emotional comfort.

There is much discussion about whether dogs have 'theory of mind' (intuitive understanding of one's own and other people's minds or mental states – including thoughts, beliefs, perceptions, knowledge, intentions, desires, and emotions – and how these mental states influence behaviour). It appears that this ability may have an evolutionary basis in dogs, inasmuch as being able to 'read' and predict the behaviour of other animals may be necessary for their survival. Selective breeding may have enhanced this aptitude.

Athough we live in the same world, there is little doubt that our dogs' understanding of it is different to ours. Their sense of smell is around 10,000 times stronger than ours, for example; they have a wider field of vision; see detail less well, and are dichromatic (they have only two colour-sensitive cone cells in their retina (yellow and blue)). In addition, their hearing is superior to ours as they can locate sound better, and hear a wider range of noises (double the frequency in the higher ranges).

... contentedly chewing a toy ...

We cannot say for certain whether dogs feel the same way emotionally as we do, although anyone who has shared their life with one can be in little doubt that they experience fear, happiness, disappointment, frustration and anger. Even though their pleasurable activities and ours may be somewhat different, we owe it to our dogs to try and look through their eyes, and provide an environment where they feel safe when alone.

There is evidence to suggest that dogs may mirror our emotions. They are certainly perceptive to our anxiety: for example, if the stress hormone cortisol spikes in us, dogs can smell this change in our body chemistry. Veterinarians anecdotally remark on the correlation of ailments between owners and pets. Behaviourists come across owners who don't feel comfortable about crating their dogs, and consequently their dogs don't like their crates. Of course, this could be a genuine dislike of the enclosed space (sometimes seen in feral dogs who have been rehomed),

... happily playing ...

... relaxing together.

It is important that our dogs learn how to relax when we're with them ... and when we're not.

a learned dislike, or simply that the owner hasn't put the necessary work into crate training, so that a dislike of the crate becomes a self-fulfilling prophecy for the dog.

But it may also possible that a dog has developed a dislike of the crate because it mirrors his owner's dislike ...

Many dogs know when their owners are about to go on holiday by picking up on their emotions and inevitable change in routine. They become clingy and anxious – perhaps they even suspected something was up when we thumbed through the holiday brochure all those months ago! Trainers have taught dogs to mimic behaviour, dog-to-person and dog-to-dog.

If dogs can emulate physical movements such as touch an object with their nose or a paw, stand on a stool, or turn in a circle, through observing the actions of a trainer, they may also observe and absorb our emotions via our facial expressions and body language, and at a neurological level through mirror neurons (a type of brain cell that fires when we carry out an action, and also when we simply watch someone else doing the same action. Mirror neurons appear to be important in interpreting facial expressions, and may play a key role in processing social behaviour, such as empathy and imitation).

If we are worried about leaving our dog, can we mask emotion so that our dog doesn't pick up on any anxiety on our part? Try getting very excited around your dog – how does he react? Next, try looking very serious – what is his response?

Observing these minute behaviour details is the key to understanding our dog's mind and emotions.

If we worry about leaving our dog on his own, it's quite possible that he will pick up on our anxiety. The skills and tools covered in this book can support our own emotional needs as well as those of our dog, since evidence suggests that our emotions appear to go 'down the lead' to our dog.

The only way to truly evaluate what your dog does when he is at home on his own is to set up a video camera to run when you are out. This can provide valuable information, and help with more accurate assessment of his level of emotional comfort.

continued page 32

How emotions may outwardly display

Negative emotions	Positive emotions	Calm emotions
Fearful and worried	Excited	Engaged in activity
Angry, frustrated and upset	Happy	Sleepy
Depressed and withdrawn	Anticipating pleasure	Relaxed and non-reactive to stimuli
↓	↓	↓
Physical manifestations	Physical manifestations	Physical manifestations
Ears back	Ears alert and listening	Eyes soft
Frequent lip-licking and appeasement behaviour. Possibly panting and taking long drinks to replace fluids lost in this way	Mouth soft, open and jowls loose. Drinking is normal	Mouth soft, open or closed Drinking is normal
Withdraws from social contact or becomes over-attached	Engages in social contact, focused and interested in surroundings	May be too relaxed or engaged in activity to move, but friendly if approached
Muscles tense and ready for flight/fight	Muscles ready for activity	Muscles relaxed
Tail tucked under	Tail erect, wagging, communicative	Tail relaxed and expressive when communicating
Lacks confidence	Confident and engaged	Sociable and friendly when approached
Aggression possible	Aggression unlikely	Aggression very unlikely
Easily startled, recovery time poor.	Less easily startled, recovery time quick	Startle response low, recovery time excellent
May become exhausted	Good level of energy	Energy calm, relaxation provides healing time

Home alone – and happy!

The Cavalier King Charles Spaniel is a toy breed historically bred as a companion dog. Of course, any dog – purebred or crossbreed – can develop a strong attachment to their owner.

Our dogs love spending time with us!

Walking our dog establishes a great connection between us ...

... playing with him builds a strong bond ...

... and simply being in each other's company binds us together in companionship. But it is this very bond that causes some dogs to be unhappy when apart from us.

Home alone – and happy!

Analysing emotions, and how these may outwardly display

Dogs are family. They share our homes, and sometimes our beds. We enjoy and encourage a close relationship: indeed, breeds such as the Cavalier King Charles Spaniel, Pekingese and Shih Tzu have been bred specifically as lap dogs. Selectively breeding our dogs in order that they provide us with close companionship comes with a price: some struggle to cope without us.

We enjoy the company of our dogs, spend quality time with them, and create emotional dependency in them; yet we also expect that they will be able to easily cope without us. If we need our dogs to spend time alone, either now or in the future, we have a responsibility to ensure their emotional well-being when they are apart from us. With the help of this book, you can teach your dog – and especially your puppy – the skills and ability to relax that he will need to enjoy his own company.

When we first bring a puppy into our home it's an exciting time for the family. But have you ever wondered how the puppy might feel about it? Suddenly taken from his mother and siblings, who have been his teachers, playmates,

We expect our puppies to cope with solitary confinement with no prior training.

comforters and constant companions for the whole of his young life, he is brought into an alien environment, and often has no contact with other dogs for several weeks until after his vaccinations are complete. It is likely that your puppy will never have been on his own until you bring him home, yet he is expected to cope with solitary confinement without difficulty. His first home-alone experience is coupled with the stress of leaving mum and his siblings, a veterinary visit and check-up, vaccinations, car travel, a new environment, novel sounds, sights and scents, and new faces: is it any wonder that some youngsters struggle to cope?

But it doesn't need to be like this. With a little forethought, education, planning and training, we can support our puppy's well-being by providing a 'safe' learning environment, in which he feels physically and emotionally secure, and gives vital guidance on the correct way to behave: chewing a bone (guided choice), or a table leg (unguided choice). Guided choices give our puppy the best possible chance of successful home-alone learning. If our puppy has an older dog for companionship, the older dog may be taken out for his walk before our puppy has completed his vaccinations. Our puppy will probably be totally unprepared to spend time alone, so his first experience of being left without company may well be a negative one. If we can prepare him by first teaching him to enjoy his own company, and spend some time apart from his companion, we are guiding our puppy to make the right choices, as well as building life-long emotional stability.

Managing emotions on departure

Preventing separation anxiety requires that we change our dog's or puppy's feelings of disappointment that we are leaving to joy and anticipation that he will get something wonderful (relaxation support in the shape of a stuffed Kong™, a bone or hidden treats to find) when we pick up the car keys or put on our shoes. Just as the keys can be the anxiety trigger, so, too, can they trigger anticipation of his receiving a great treat. But before making the connection between a treat or favourite toy and our departure, it is essential to teach our dog or puppy to first enjoy the treat in our company.

If the relaxation support (radio, treat, stuffed Kong™ etc) is given only when we leave the house, it is easy for our

continued page 38

An older dog and a puppy in the same household often form a strong bond.

Home alone – and happy!

If our puppy has not had all of his vaccinations and his 'best friend' is taken for a walk, how might our pup feel about his first home-alone experience?

A relaxed home-alone dog may find toys comforting whilst we are away.

Gauging and managing our dog's emotions on our departure is key to managing his contentment when he is alone.

Home alone – and happy!

Left: Accustom your puppy to eating from his Kong™ whilst you are still around, so that he anticipates this activity when you leave him alone with it.

Above: Eating from a Kong™ will ensure his dinner lasts longer, and slows down those dogs who eat quickly. It also offers a puzzle-solving opportunity – and is tiring!

Home alone – and happy!

dog or puppy to become sensitised to this, and the very thing we are hoping will comfort and occupy him whilst we are away, actually becomes a trigger for his anxiety! For instance, if we give our dog or puppy an interactive toy when we leave the house without first teaching him how to enjoy it, the toy can become associated with our leaving. Our dog then becomes anxious on seeing the toy; his stress level rises, and his appetite may diminish. Food is the last thing on our minds when we are scared, and many dogs with moderate to severe separation anxiety will not eat when alone – one of the signs that they need support.

Establishing solid foundations

Invest quality time in teaching your dog or puppy to enjoy his food, and/or home comfort aids whilst you are available to provide emotional support, gradually weaning yourself away as his enjoyment of these grows, and he becomes less reliant on you for support. In any type of training, building robust and solid foundations are the most important part, and cannot be rushed. Without these, behaviour crumbles under pressure, and dogs often revert to previous learned behaviours: disappointment, barking, chewing, etc, and a dog without the necessary coping skills will dread your departure well in advance of it actually happening.

Before leaving him alone, make sure that your dog or puppy is happy to enjoy the comforting treats you have provided him with while you are around and about the house. Acclimatise him to the fact that you are not always there by leaving him for slightly longer periods, beginning with just a few seconds and building to minutes.

Incorporate training into your daily routine: for instance, time your dog's breakfast with making a cup of tea in another room, building to when you have a shower each morning. Increase the amount of time he spends alone so gradually that your dog doesn't notice he is being left for longer. When he is engaged in one of the activities described in this book, pick up and put down your keys, and carry out other activities such as putting on your shoes, without actually leaving, so that these events become mundane.

Building anticipation

When our dog or puppy begins to anticipate that our departure means he receives a treat, we're on the right track!

Preparing your dog's food shortly before you leave the house, and putting it to one side whilst you put on your coat and shoes, can build anticipation in your dog and increase the appeal and value of the food.

38

If happy and distracted, our dog or puppy will be aware that we have left the room – and may even look up – but should quickly return his attention to his treat. If he whines, ignore this unless he is becoming distressed. If you go to him straight away when this happens, he may simply learn that whining or barking initiates your return. If he does become distressed, it's possible you have gone too quickly, and more time needs to be spent teaching him to enjoy his comfort aids. Before you can leave him for even a few seconds, a strong feeling of enjoyment (positive reinforcement) must be established and associated with these treats.

REMEMBER!
Anticipation is pleasurable! When your dog or puppy has learned to eagerly anticipate his food-dispensing toy, prepare it and leave it out of reach but in view while you prepare to leave. He will be full of anticipation of the pleasure this will bring, but do not make him wait more than a few minutes for his treat, otherwise this could tip over into frustration.

3: Building independence

Mealtime training: feed him and leave him!

We can use mealtimes as opportunitues to teach our dog or puppy the skills he will need to be happy when alone. Building regular training sessions around each mealtime doesn't require a lot of time-investment; just a little planning and a few extra minutes to fill a Kong™ or activity toy, and mealtime training sessions also help avoid weight-gain. A 14-week-old puppy on three meals a day can be left alone three times during the day, providing at least 21 opportunities each week for home-alone practise.

By teaching our dog these essential skills we are future-proofing his emotional needs. Of course, we cannot possibly protect our dog from every eventuality when

Begin to build your puppy's independence by allowing him to play on his own with a toy while you supervise from a distance.

Feed him and leave him! Encourage your puppy to eat all of his meals from an interactive toy, which will allow you to train home-alone skills around his meal times.

Home alone – and happy!

A new baby can mean that our dog spends more time alone. If not prepared for these life changes, he may become anxious about being left.

we're not with him – a thunderstorm may brew, a picture may drop off the wall and startle him, or a jet might fly low overhead. However, if our dog feels secure when alone, and he has acquired strong foundations of self-reliance and self-confidence, he is more likely to have the coping strategies necessary to deal with the trials and tribulations that life throws at him, especially when we're not with him.

Family circumstances change, too. The puppy who goes everywhere with a newly-married couple, suddenly finds himself at home alone more often because their new baby is now their priority. The stay-at-home mum who re-homes a rescue dog decides to return to part-time work when family commitments allow, meaning that her dog now spends more time alone. Someone who works from home is offered outside employment, so their dog then stays home alone for several hours a day. If we can prepare our dog for a time when our personal circumstances may change, and he may need to be at home on his own more often, he will be much happier should this happen.

Eating for pleasure

For most dogs, eating is one of life's great pleasures, and can be used to help promote independence and solitary contentment. Many dogs are fed once or twice a day from a food bowl in the kitchen when their owners are nearby, and mealtimes are usually shortlived. Eating times are, however, a perfect opportunity for teaching our dog to enjoy his own company.

Put away his food bowl, and be creative with your dog's mealtimes! Teach your dog or puppy that the trade-off

continued page 47

Foraging for hidden treats engages our dog's natural instincts, and is hard work! A few treats scattered in the garden can engage and calm a puppy, leaving him happily tired.

A very simple game is to hide treats between the folds of a blanket or towel that your dog can forage for when you leave him.

SnuffleMats™ are strips of knotted fabric in which treats can be hidden: they also provide a stimulating way of feeding your dog his kibble dinner.

Nosework is mentally and physically tiring. The SnuffleMat™ also doubles up as a comfy pillow!

for being alone is the pleasure of eating, and start this from the day your dog or puppy comes to live with you, if his appetite is good. (Some puppies are reluctant to eat for a few days, often brought about by the stress of leaving mum and siblings, in which case wait for your puppy's appetite to pick up before starting this regime.) One simple activity is to scatter your dog's kibble, or some treats, around the garden for him to forage on his own. A favourite canine pastime is scenting, so rummaging around in the grass for food is likely to be very enjoyable.

Foraging engages the seeking circuit, the part of the canine brain that makes exploring the environment – especially for food – exciting and pleasurable. The seeking system is essential for survival, and harnesses a dog's natural instincts. Start by scattering the food in a small area of the garden to make it easy for him to find, gradually increasing the difficulty by spreading the food over a larger area, allowing him to spend longer happily foraging while you are in the house. (Never leave your dog alone in the garden when out of the house. You will have no control over what

Giving our dog a food-dispensing toy before he is left alone allows us to pair our departure with something pleasant for him, and there are a great many different interactive food dispensing toys on the market. Treats can be hidden in an activity ball to chase …

... or placed in an open-ended plastic bottle, making accessing the treats a fun, puzzle-solving game.

The challenge is to tip the bottle on its end to allow the treats to drop out!

he does, and he may practise unwanted behaviour, such as barking at passersby, or chasing cars up and down the fence.) Food can also be hidden in the house for him to search for and find whilst you're out.

There are many food-dispensing toys on the market that provide mentally stimulating and puzzle-solving ways to distribute food over a period of time, and our dog or puppy can be left alone with these once he has understood how to use them. If we leave him alone with a food-dispensing toy before he has learned how to extract the food, the toy could become a source of frustration – exactly the emotion we want to prevent. Teaching him how to use the toy whilst we

are around, and allowing time for him to properly learn this, will ensure our dog is excited about the prospect of playing with the toy when we leave him alone with it.

Food ingredients

Carefullly consider the ingredients of your dog's food. Poor quality foods are hard to digest, and can make dogs feel quite uncomfortable. Food labelling can be quite confusing, but, in general, actively avoid any kind of food or treat where the origin is uncertain. Avoid meat and animal derivatives; by-products, and pulp, since the nutritional value of these ingredients is doubtful. Rawhide chews may be indigestible and have little nutritional value – and there is also a very real risk of your dog (and especially a puppy) choking on these, as they can become lodged in the throat. Some commercial foods and treats are high in sugars, colours, derivatives and preservatives, which can make dogs 'hyper,' and have a

A food-dispensing Kong™ wobbler 'brain teaser' toy can make meal times last longer, and keep our dog occupied whilst we're away from him.

detrimental effect on behaviour and their ability to relax, just like a child on a sugar 'high'. You could consider baking your own nutritious treats.

Check labels to ensure you feed only high quality foods, such as a balanced raw diet or good quality tinned food or kibble that clearly lists all the ingredients. Good

quality tinned foods or raw diets have the additional advantage that they are easy to stuff into a Kong™ and smeared around dog bowls with special ridges which are designed to slow the rate at which a dog eats his food. On a hot summer's day your dog may enjoy a stuffed Kong™ that has been frozen (which also means it will take him longer to

Home alone – and happy!

eat the food), or a bowl containing frozen stock with biscuits or his kibble dinner dropped in. Or what about leaving your dog with a Kong™ stuffed with yoghurt and frozen? An 'ice cream' on a hot day – lovely!

Fussy eaters

Some dogs are naturally fussy eaters. There may be constitutional reasons for this, it could be learned behaviour, or your dog may even self-regulate his food intake. Although most dogs, if given the choice, will over-eat, anxiety suppresses appetite, and a dog who anticipates your departure may not feel inclined to eat whilst you are away. Smaller breeds, on the whole, tend to be less greedy and fussier eaters.

One reason for his lack of appetite could simply be that he is full, and his food (including treats) is supplying him with more calories than he needs. If your dog is a fussy eater, always speak to your vet in the first instance, as there could be an underlying medical cause, such as reflux or colitis, with associative pain when eating. Some fussy eaters do better on a homemade diet, and warming his food can help stimulate

It's easy to bake appetising homemade biscuits for your dog, which can be given as a treat when he's left alone. Online dog 'delis' and some pet stores also stock wholesome treats and biscuits that are both good to eat, and good for him!

the appetite; if you are in any doubt, speak to a professional to ensure your dog is receiving a balanced diet

If your dog is missing meals but is otherwise healthy, energetic, and maintaining good body weight, and an underlying medical cause has been ruled out, he may simply be getting too much food. Metabolism varies from dog-to-dog, as it does from person-to-person, and other factors come into play such as amount of exercise. Try offering your dog one meal a day, perhaps, to stimulate his appetite and reward his hunger, so building the pleasurable association that anticipating his dinner satisfies his hunger. Ensure his eating area is a quiet place away from other animals and the hustle and bustle of the household, and leave him alone to eat to prevent the possibility that he may be picking up on your anxiety about his eating habits. A puppy whose rapid growth phase is coming to an end at around 5-7 months of age will sometimes have a reduced appetite, and require less food, and hormones can negatively influence appetite when adolescent males experience hormonal surges, and bitches come into heat (oestrus).

Once your dog is looking forward to mealtimes, and happily eating his dinner, begin his home-alone training. If his appetite remains poor, however, seek advice from a professional without delay.

Feeding raw bones and bone safety

Raw marrow or knuckle bones have the advantage of keeping our dog or puppy occupied for a long time. They provide an upper body work-out as well as mental stimulation, plus clean teeth, build strong shoulder muscles, and promote paw dexterity. As previously noted, chewing releases 'feel-good' chemicals which encourage our dog to be content when alone.

To avoid unwanted weight-gain, as the marrow in raw bones is both highly nutritious and very rich, you may wish to feed these in place of a meal, one or two days a week. An old towel placed over your dog's bed will help to keep it clean, and if your dog is happy to use a crate, feed him the bone there.

Please be aware that some dogs may guard raw bones because they are such a high value resource, and some animals may be fearful of the bone being taken from them. A number of factors may also increase a tendency to guard food, such as a genetic predisposition combined with

Only begin your dog's home-alone training when you're confident that he is enjoying his mealtimes.

Providing sensible precautions are taken, nothing can beat a good gnaw on a raw marrow bone to stimulate your dog's mind, exercise his body, and feed his soul!

environmental causes: for example, someone walking too closely past his food bowl, children interacting with a dog whilst he is eating, or another family pet close by.

Taking your dog's raw bone from him to show him 'who's boss' is not a solution, because it assumes that your dog understands why his valued resource is being removed (which he obviously does not), and may cause anxiety to build over time. It's also an unkind way to treat him. Imagine a scenario where you are enjoying your Sunday roast, and someone removes the plate from under your nose. While you might tolerate this, you will probably not be best pleased, and the urge to protect your dinner may grow to the extent that you take steps to prevent its removal. If your dog growls when you approach his bone, dinner or prized possession, and you, understandably, back away, he may learn that guarding behaviour is effective: you back

away and he keeps his resource. Behaviours that your dog considers to be successful will very likely be repeated.

Your dog must have a safe, quiet place where he can eat and drink undisturbed, and, if crate training, it is recommended that you feed him all of his meals in his crate, so that he learns to associate this area with a pleasurable activity. If there are young children in the household, feed your dog in a covered crate to prevent his being disturbed by toddlers wanting to stroke or play with him, say, while he is eating. He may even think that the child wants to steal his food, and so begin guarding it.

We can teach our dog that hands have positive

Home alone – and happy!

connations by hand-feeding him (using a flat 'pony-feeding' hand), holding a Kong™ for him to eat from, or placing several, high value morsels in his food bowl when he is eating.

Having someone bring us more gravy, a second helping of Yorkshire pud, or a chocolate dessert will be infinitely more welcome than the individual who takes away our dinner ...

Should we give a raw bone to a dog who guards, or does this allow him to practise guarding? There are few interactive feeding toys that can keep a dog occupied for the same length of time as can a raw bone. Some dogs who are not particularly food-motivated will enjoy a bone whilst we are out, and a dog who has chewed a raw bone for an hour or more will be content and relaxed. Providing you can feed your dog a bone in a safe place, such as a covered crate, away from other household pets; can lure him away from the bone with a very tasty morsel such as fresh chicken, and are aware of potential guarding issues, on balance, the benefits of doing so outweigh the risks – just be mindful about how, where and when you feed them. (Don't try and tempt your dog away from a raw bone that has just been given to him: wait until he has had it for some time, when he is more likely to relinquish it.)

Our dogs live in an unnatural environment, where we restrict many of their instinctive behaviours. They cannot watch television or read a good novel for pleasure and relaxation like us, and gnawing on a raw bone can make up for this by acting as an excellent stress-reliever. Raw bones provide stimulation for the mind, exercise for the body, and food for the soul.

REMEMBER!

Please NEVER feed cooked bones, as these are very dangerous for your dog. Cooking changes the bone's composition, and it can become brittle and splinter, causing serious injury, and even death. Do not leave your dog alone with a raw bone for long periods of time, either, and, when feeding raw bones for the first time, stay with your dog. Only leave him once you are happy it is safe to do so, checking him at regular intervals.

Teaching 'settle on bed'
Teaching our dog a settle cue on his bed (or in his crate) can help instil independence. It's a position our dog assumes whilst awaiting his comfort treat, full of anticipation! A settle is when our dog lies down with his weight on one hip, and is a position of relaxation where he can get comfortable for a few minutes or longer, and becomes associated with being left. Ask your dog to settle on his bed, and give him a stuffed Kong,™ for instance, when you leave him, but remember, it's not a 'down-stay,' so he can move off his bed when he has finished and you are gone.

To teach a settle, first train a 'down' position, then entice him with food onto one hip and into a 'settle.' (A down position has the weight evenly distributed on both hips, like a sphinx.) Teach your dog 'down' by holding food in front of his nose, and gradually lowering your hand to the floor. Hold the treat with your fingertips pinched together when enticing him, and offer the food from your open palm once he has assumed the correct position, so that he understands (from your hand position) when he can take the treat.

Once in the down position, entice your dog onto one hip by placing the food on the floor behind his front legs, next to his tummy. Crouch down and feed him every few seconds to maintain the position. Gradually stand, but still maintain a high rate of feeding. When his 'settle' is reliable with you next to him, and you can stand up, take a small step away from your dog. Remember to keep feeding him on the floor by his tummy every few seconds.

Your goal is to gradually extend the time he spends in the settle positon, whilst you move away for longer periods of time, and you can eventually leave him on his bed whilst you prepare his Kong™ or activity toy. For your dog, going to his bed in this way should be filled with pleasurable anticipation and excitement of what he will receive whilst you are out of the house, and is a great attitude to teach him.

Building independence
Divide your dog's meal into three training portions, or four if your puppy is up to 12 weeks of age. Choose an activity to do from the examples in column 2 (My activity) of the table on page 56 that will take between one and fifteen minutes to achieve, starting with the least time.

Whilst carrying out your chosen activity, give your dog his *Home alone – and happy!* activity shown in column 3 (My dog's activity).

continued page 57

Training the 'down' position with food. (Courtesy Sachiko Eubanks)

Feed in the 'down' position. If your dog should begin to rise as he anticipates the treat, feed him on the floor between his paws near to his chest to maintain downward focus. (Courtesy Sachiko Eubanks)

Change your enticing hand to a hand signal for 'down.' Feed your dog in this position. (Courtesy Sachiko Eubanks)

Encourage your dog into a 'settle' position with his weight on one hip by using the treat to draw his head round to the side, then feeding by his tummy. (Courtesy Sachiko Eubanks)

Regularly practise the 'settle' position with your dog, and he will soon learn to anticipate good things when you ask him to wait on his bed! When releasing your dog from this position, give a verbal cue, such as 'okay,' as you throw a morsel of food for him to chase. (CourtesySachiko Eubanks)

Actively encourage your dog to settle in another room, away from you, whilst you are at home. If he can do this, he is more likely to cope when you're apart.

Home alone – and happy!

My dog's meal:	My activity (choose one)	My dog's activity	Evaluation (see page 57)
Breakfast	1 Picking up car keys & leaving the room 2 Brushing teeth 3 Having a shower 4 Getting dressed	• Eating from a stuffed Kong™ while wearing a bandana sprayed with calming scent	Length of time my dog was left . . . . . How did he cope? A He was quiet whilst eating and afterwards, and calm on my return B He was quiet until he finished eating C He did not eat, or ate little, and was vocal
Lunchtime (snack)	5 Preparing lunch 6 Doing a household chore 7 Putting on outdoor shoes & leaving room 8 Popping out to the car	• Eating from a Kong™ wobbler or food-dispensing toy while relaxing music playing	Length of time my dog was left . . . . . How did he cope? A He was quiet whilst eating and afterwards, and calm on my return B He was quiet until he finished eating C He did not eat, or ate little, and was vocal
Tea-time	9 Making dinner 10 Making a phone call 11 Doing a household chore upstairs 12 Putting on outdoor coat & leaving room	• Chewing on a raw marrow bone or commercial chew	Length of time my dog was left . . . . . How did he cope? A He was quiet whilst eating and afterwards, and calm on my return B He was quiet until he finished eating C He did not eat, or ate little, and was vocal
Evening (optional)	13 Preparing supper 14 Watching television 15 Getting ready for bed 16 Preparing a nightcap	• Foraging for treats in the garden/house, hidden biscuits folded in a blanket or in a box filled with shredded paper	Length of time my dog was left . . . . . How did he cope? A He was quiet whilst eating and afterwards, and calm on my return B He was quiet until he finished eating C He did not eat, or ate little, and was vocal

Carry out the suggested exercises daily, and monitor and assess your dog's mental state against the evaluation criteria. Only increase the time he is left alone when you are certain that he is looking forward to his treat, and is quiet on your return.

Vary the activities you give your dog, and the time you are away from him, very gradually building up the amount of time he is left alone.

Record your findings over a four-week period on the chart provided on page 58 (we suggest you make a photocopy for each of the four weeks), according to the evaluation criteria below, and note any significant behaviours.

Be diligent in completing your records, as they serve as an excellent reminder of how far your training has progressed.

Evaluation

A

He was quiet whilst eating and afterwards, and calm on my return

This is excellent! You are well on the way to having a dog who is content to stay home alone. If your dog is sufficiently occupied with his chosen activity whilst you are in the house, this is a great building block for his coping when alone.

B

He was quiet until he finished eating

Give your dog an activity that lasts longer, or leave him for a shorter period of time, so that you return before he has finished his activity. Spend more time with him when he is learning to enjoy his food. These are good foundation skills but need to be built on.

C

He did not eat, or ate little, and was vocal

Your dog is having a hard time coping alone, and it's likely he is anxious when you are out of the house. More work is necessary.

Stay with him until he learns to enjoy his chosen activity in your company, before leaving him for very short periods of time. Start with a few seconds, building to a minute, then 2, 3, 5 minutes, etc.

The first few minutes when you leave him alone are critical, and will be a good indication of his overall coping strategies, so practise lots of bed 'settles', when you can frequently reward your dog for staying calmly on his bed.

Home alone – and happy!

Week number	My activity	My dog's activity	Time my dog was left alone for	Evaluation (see notes page 57)		
Breakfast				A ☐	B ☐	C ☐
Lunchtime (snack)				A ☐	B ☐	C ☐
Tea-time				A ☐	B ☐	C ☐
Evening (optional)				A ☐	B ☐	C ☐

Start date ....

Day of the week ....

NOTES

....
....
....
....
....

A dog who is able to cope within the first few minutes of being left alone is likely to be comfortable for the duration. An anxious dog will have anxiety building well before we leave the house; almost from the moment we think about doing so! Dogs read our body language like a book, and sense our emotions.

4: Managing the environment

Careful management of our dog's environment can promote and support his ability to relax. If your dog or puppy is left in a busy area where there is constant visual and auditory stimulation, it makes relaxation harder for him. Imagine being at a yoga class, say, with people walking past the large, picture windows every few minutes, busy traffic on a nearby road, and lots of movement and noise: not exactly an ideal situation in which to relax!

Beds and sleep quality

Consider where in your home your dog sleeps. An area that is too noisy – such as in the lounge when the TV is on constantly, somewhere that is too hot, cold or draughty, or too open (not den-like) – are all unsuitable bed positions, which will mean that your dog is deprived of emotional comfort, and is easily woken, startled, or aroused.

Size and type of bed are also important. While some

Deep, relaxing sleep is essential for our dog's overall health and wellbeing. Ensure his bed is comfortable, and is the right size.

Dogs often have favourite resting places – even if they may seem rather unlikely to us ...

dogs like to curl up, others prefer to sleep lying flat on their side: a position that allows complete relaxation for deep sleep. Is your dog's bed big enough? A bed that is too small can cause back problems. Would your terrier appreciate a sleeping bag-style tunnel bed? Does your elderly dog need a firm, raised bed?

Keep your dog's bedding clean: Vet Bed™ is a product that can be very easily washed and tumble-dried. Some dogs love to 'nest,' and will dig and paw their beds until they are comfortable, so perhaps provide a loose blanket on top of his bed for this purpose.

Deep, relaxing sleep that repairs, rejuvenates and increases dissipation of stress hormones is essential for health and vitality. A dog who is constantly aroused and on the look-out for threats may produce more stress hormones such as adrenaline and cortisol, which can have a negative impact on health, deprive him of good quality sleep, and, cause him to become more reactive and less likely to settle when home alone. Consider the effect that light pollution has on your dog's evening and night-time sleep quality, too. If his sleeping area is too bright at night, less efficient production of melatonin – an important hormone that

... and some find the gentle pressure of resting their chin comforting: especially if this is on a favourite toy.

Home alone – and happy!

helps regulate the body clock – can be the result. Light from electrical appliances, especially in the kitchen or utility, can considerably illuminate a room where your dog's bed might be situated.

Dogs spend a considerable amount of their day lying down, dozing, resting and sleeping, and this is the relaxed Fido we are aiming for when we are out.

Room temperature
Consider the overall temperature of the room or area where your dog will be whilst you are out. The normal body temperature of a dog is marginally higher than ours at between 38.2 and 39.3°C (101 and 102.5°F), and dogs generally like the environment to be a little cooler than we do. Do not leave your dog in a conservatory when you go out during the summer months, as he could very easily become over-heated, with attendant risk to health, and even death. Dogs often have a favourite sleeping place at home, although they will move around. Wherever you leave him, make sure your dog can can find a cooler or warmer spot to maintain his comfort, and always ensure there is clean drinking water available.

Ensure your dog is a comfortable temperature. If he has been for a walk in the rain, or gone swimming, put an absorbent drying coat on him to prevent him getting chilled.

The body temperature of our dogs is higher than ours. Ensure your dog has somewhere cool to rest in summer when you are out. Beware of conservatories becoming too hot – especially in summer!

Reducing arousal

Dogs or puppies who watch from a window when we are out may become too aroused to relax, by becoming over-stimulated by their environment. For instance, some dogs may hone their instinctive predatory motor pattern of eye-stalk-chase on cars driving past, 'eye' people coming and going, bark at birds through the window, or guard the house, barking at passersby who then, obligingly, disappear! Small terriers can sometimes be found standing on the top cushions of a sofa (as this usually gives an excellent vantage point), guarding their territory and barking at passersby. These types of behaviours which, in our dog's eyes, lead

Home alone – and happy!

to a successful outcome, are likely to be repeated, but the high arousal levels required to maintain vigilance are not conducive to relaxation.

Some breeds are very sensitive to their environment, which a working Border Collie, for example, needs to be in order to do his job. Excellent eyesight to detect the slightest movement of the flock, sensitive hearing to listen and respond to verbal instructions from the farmer, and an awareness of small changes in his environment are all prerequisites. While these attributes are perfect for his job of herding sheep in rural conditions, they may make him more anxious about changes in his home domain, acutely aware of sudden and loud noises and raised voices, and less able to relax in stimulating environments. If his bed is by a window, through which he can see cars driving past on a busy road, he may find it hard to relax there, for example.

Consider your dog's breed history when assessing his environment, and, if necessary, manage the environment: limit what he can see by shutting a door, restrict access to windows, close a blind or put up a net curtain. Create a 'yoga class' environment for your dog, one that is calm, relaxed, quiet, low-lit and safe, without stimulation. If you use a crate

Perhaps partly close blinds to reduce what your dog can see through windows. Minimising visual stimulation will enable him to relax more deeply.

for your dog, cover it with a blanket to help him feel more secure. After all, we close our bedroom curtains and turn out the light when we want to sleep.

Movement restriction

Don't let your dog or puppy follow you everywhere in the house. Stair gates are perfect for managing the environment and restricting access, and one placed at the bottom of stairs, say, will prevent your dog or puppy from following you upstairs. This method has the added advantage of controlling his whereabouts without the need for constant verbal intervention, and can be very useful for house-training. Stair gates also teach our dog that there are times when we are both in the house, but need to be apart: a key skill our dog needs to learn in order to be comfortable when alone. And because he will be able to see through the struts, a stair gate can make a dog or puppy feel less isolated than would closing a door.

The crate: your dog's bedroom

Spend some time teaching your dog or puppy to enjoy his crate. Some dogs and puppies need to learn to relax in their crate first, and also eat there, especially if the crate has previously been used as a 'time out' aid. Leave the crate door open during the day so that your dog or puppy can wander in and out of his own volition, and cover it with a blanket to make it feel more secure and den-like.

Dogs will often seek out secure places to rest, such as under a table, in a corner, or against a wall, and crates are healthy alternatives, providing places of safety to avoid something they dislike, and secure areas in which to relax and eat their food.

Leave your dog's crate door open when you go out, so that he has the option of eating or sleeping in there, and please don't shut your dog or puppy in a crate for long periods of time. Doing so is cruel as it restricts his freedom and isolates him, depriving him of the vital social interaction he needs. However, feeding him a Kong™ there saves possible mess on the floor, safely separates dogs in a multi-dog environment, and provides a secure place for your dog to enjoy his treat.

Providing they are used appropriately, crates can be an excellent management tool, and leaving the door open gives your dog the choice of whether or not to enter. If you

Stair gates are useful for restricting access, and also as a way to keep him safe. They teach our dog that some areas are out of bounds, and that he can't always be with us, although he may still be able to see us.

Leave open the door of his crate during the day to give him the choice of being in or out of it. Many dogs actively choose to rest in their crate.

Home alone – and happy!

feel you need to shut your dog in a crate because he is being destructive when left alone, please seek professional help without delay.

Covering all or part of our dog's crate with a blanket can help make him feel secure and cosy, as well as keep him warm.

Closing the crate door at mealtimes, for just a few minutes' confinement, is a useful practice, as there may be times when it's necessary to safely confine your dog: if he has an operation, say, if visitors, such as very young children, need to be distanced from him, or if a workman is in and out of the house. Once your dog or puppy is comfortable eating from a Kong,™ leave him alone in his crate with it, and remember to use his meal to build home-alone skills – a puppy on three or four meals a day can be left alone each time he eats.

An alternative to using a crate is to place your dog or puppy behind a stair gate or in a puppy pen, which is much less isolating and stressful than closing a door on him.

The sound of music
Some musical melodies can positively influence your dog's

Feeding our dog his Kong™ in his crate can prevent a messy floor ... and teach him that his crate is a great place to be!

Home alone – and happy!

home alone relaxation. Dogs respond to the same sort of calming music that we do, and listening with our dogs can connect us in the moment.

In order to make a significant difference, it appears that playing the right type of music is important, and even classical music can vary in the effect it has, from calming to arousing. Vibrations in the lower register rather than treble have a more tranquil effect on the nervous system, with suitable music helping to minimise environmental stress, cut out unpleasant and exhausting background noise, and enhance our dog's overall well-being. In one study in a shelter environment, heavy metal music was shown to agitate dogs, while classical music had a calming effect, and neither conversation nor pop music appeared to affect a

dog's behaviour.[3] Do be mindful of not simply playing songs or radio stations that *you* like when you leave your dog alone.

Simple music (one or two instruments) played at around 60 beats per minute, the same as a resting heart rate, can help calm your dog's nervous system and train his mind to become more peaceful. Consider purchasing music that has been specifically composed for animals, and which encourages passive rather than active listening. Always allow your dog to move away from the sound source if he wishes to, and don't play it continuously.

For music to be most effective, begin playing suitable tracks when your dog or puppy is already happy and content: perhaps when eating his dinner from his interactive

The right kind of music can have a calming influence.

3 Graham, L, Wells, D L, Hepper, P G (2002). The Influence of Auditory Stimulation on the Behaviour of Dogs Housed in a Rescue Shelter. Animal Welfare 11 (2002): pp385-393. Summary available from http://www.throughadogsear.com/research/.

In our world of constant white noise, remember that our dog, with his sensitive hearing, will benefit from some quiet time – just as we do.

food-dispensing toy. Be aware that if you only turn on the radio 'for company' when you leave the house, this, in itself, could trigger anxiety as your dog will come to associate your leaving with the radio being turned on. Switch on the music well before you leave the house, perhaps 30 minutes beforehand, to help it take effect. If you leave your television on or, indeed, play TV especially designed for dogs, do ensure that he will not bark at other animals he might see, and that this strategy is having a relaxing rather than arousing effect.

Dogs have an acute sense of hearing, and, because they hear almost twice as well as us, will often prick up their ears at sounds that we cannot hear, so make sure that the music you play is at a suitable volume for him. We are bombarded with noise pollution these days, so maybe our dog would enjoy some time without the radio on, or music being played, and perhaps he might thank us for a bit of peace and quiet on occasions when a normally noisy household is hushed. We live in a busy world of constant auditory stimulation where the sound of silence is rare: we should be careful not to inflict too much noise on our dogs.

Scent

Imagine what it would be like to live in a world dominated by scent, where all around us was a potpourri of individual odours, each with their own meaning. Scent is a very powerful sensory tool for our dogs which can evoke strong emotions. Possessing a sense of smell that is approximately 10,000 times more sensitive than ours, a dog can detect substances at concentrations of up to one hundred million times lower than we can perceive.

The olfactory bulb, where information about odours is processed in the brain, is about the size of a plum in our dogs, compared to the size of a raisin in us. Dogs also have a second sensory system for olfaction called the vomeronasal organ, and we may see our dog licking the air, mouth-smacking and tongue-flicking, actions which allow scent particles to be captured by the tongue, opening ducts to allow pheromones to access the vomeronasal organ. Scent memory lasts for many years – if not for a lifetime – and, as we might recognise an old friend by their facial features, our dog might recognise them by their scent.

Certain scents can make our dogs feel more secure, and produce a calming effect. There are a number of products available which are designed to calm our dog, from species-specific, man-made odours to herbal diffusers and essential oils. Adaptil™ is available as a diffuser and spray, and produces pheromones that mimic a bitch suckling her puppies. Pet Remedy™ (diffuser and spray) works differently by enhancing the production of GABA (Gamma Amino Butyric Acid), a calming neurotransmitter. It contains a unique blend of naturally de-stressing and calming Valerian essential oil blended with vetiver, sweet basil and sage, and works by 'tricking' nerve cells that are fired by adrenaline into believing they are receiving a calming message from the brain. Adrenaline is a hormone secreted by the adrenal glands in response to stressful situations, and, after trauma, can stay in the system for several hours before returning to normal, leaving an animal more vulnerable to the stressful effects caused by being alone.

For most dogs and puppies, moving to a new home, when familiar surroundings change, can be a stressful event, and even if the animal is not displaying overt signs of anxiety, it is worth considering the use of calming scents as part of an overall relaxation protocol when moving home or welcoming a new dog or puppy into your home.

Home alone – and happy!

This dog's bed has been carefully placed near a calming plug-in diffuser. Pairing calming scent with the pleasure of eating promotes and supports relaxation.

Some dogs find the pressure of a coat or t-shirt calming, and these can be sprayed with a relaxing scent, too, for additional support. However, in warm weather, be careful that your dog does not overheat when wearing additional clothing.

Essential oils such as lavender and chamomile have been shown to induce relaxation in shelter dogs, who also vocalised less when exposed to these odours.[4] The calmer your dog or puppy, the more likely he is to relax and enjoy his own company, and the more successful your training will be, as stress inhibits learning. Dogs 'see' through their nose, so scent is an important sense which, when used appropriately, is an essential part of your toolbox.

Some dogs find the pressure of clothing comforting. American animal behaviour expert Dr Temple Grandin's autism gives her particularly good insight into animal behaviour, because autism and sensory overload often go hand-in-hand. Dr Grandin built herself a 'squeeze' machine that helped her feel more secure, as she found that pressure contact helps release calming oxytocin, a neurotransmitter implicated in mother-infant bonding. The canine equivalent of this is a t-shirt or snug-fitting coat.

Some dogs will appreciate being left with an item of clothing that contains your comforting scent: most of the objects they 'steal' around the house are not new items but older, more heavily-scented ones such as slippers or socks. Some calming products can be sprayed onto a t-shirt to increase their effectiveness, or on a bandana placed around your dog's neck.

4 Graham, L, Wells, D L, Hepper, P G (2004). The Influence of Olfactory Stimulation on the Behaviour of Dogs Housed in a Rescue Shelter. *Applied Animal Behaviour Science Journal* (May 2005), vol 91, issues 1-2, pp143-153.

Spray your dog's bandana with a calming scent ...

... and put it on him before you leave the house. A puppy may chew a bandana so a t-shirt, or spraying his bedding, may be safer options.

5: Home alone and happy toolbox

There are many aids and tools on the market that can support our dog's training in home-alone skills, and the better engaged in and distracted by his enjoyable activities he is, the more pleasurable and comfortable will be his time alone.

REMEMBER!
Pay careful attention to creating a calming environment when you introduce the interactive tools: regard the training and aids you use as a complementary package. If your dog is reluctant to eat, consider introducing other aids or

Even if our dog will not eat when home alone we can still enrich his environment by introducing other calming tools, such as spraying calming scent on a bandana, playing relaxing music, and providing favourite toys.

An object – such as a brightly-coloured towel – very visibly placed in the home when you are going out without your dog, can communicate to him your intention, and help prevent the emotional rollercoaster of expectation, when he sees you making preparations to leave, and disappointment when he realises he is not going, too.

Home alone – and happy!

approaches first to support relaxation, and in time your dog may feel calm enough to enjoy food when he is alone.

Environmental cues

Consider placing a visual cue (object) somewhere to let your dog know when you are going out, and that he isn't coming with you on this occasion.

Once he is happy being left alone, place an easily visible object, such as a brightly-coloured towel, on the door handle a few minutes before you leave, to signal that he is staying at home. Remove it when you come home (placing it on the door handle acts as a good reminder to do this).

This strategy can help avoid the peaks and troughs of excitement when our dog thinks he is going out because we are making preparations to do so, followed by the disappointment of realising that he isn't.

Ensure that the object you use contrasts well with the background: a dog's colour vision is similar to red-green colour blindness in a person, and he has difficulty distinguishing between green, yellow and red.

Key training reminder

Don't leave your dog alone with a food-dispensing toy until you are certain he has learned to enjoy it.

Firstly, teach him the skills he needs to know how to use the toy; secondly, leave him with the activity whilst you are still home, and thirdly, leave him with the activity when he is home alone. It is vital that your dog is taught to enjoy his home-alone aids in this order.

Always ensure that your dog has the skills required to reach the food in food-dispensing toys before you leave him alone, as being unable to do this may add to his frustration and anxiety.

Your toolbox will consist of –

• Environment
Create a calming, non-stimulating environment by:
✔ Playing music specifically composed for animals
✔ Closing blinds or curtains to reduce visual stimulation
✔ Supporting relaxation with calming scents
✔ Ensuring ambient room temperature
✔ Proving comfortable, appropriately-sized bed or crate

• Distraction
Provide interactive toys, and find other ways to gradually dispense your dog's food, with, for example:
✔ Interactive, puzzle-solving feeding dispensers
✔ Raw marrow bones
✔ Frozen Kong,™ and 'smoothies' in summer
✔ Hide-and-seek games around the house, in blankets or SnuffleMats™

• Interaction
Using your dog's love of new things, and play, supply:
✔ Toys reserved only for when your dog is home alone

• Assessment
Regularly assess your dog's level of comfort by:
✔ Using the worksheets provided
✔ Remembering that your dog's emotional comfort can fluctuate according to circumstances
✔ Encouraging all family members to participate in his training

REMEMBER!

Even if you feel that your dog has become comfortable with being left alone, and perhaps doesn't need his usual aids, don't stop enriching his environment, as this may result in his regressing, and becoming anxious. Keep on doing whatever it is that supports him.

Kongs™ and other interactive toys

A Kong™ is an essential tool for teaching our dog how to be happy alone. The first step is to teach our dog or puppy to enjoy eating from his Kong.™ Kibble can be first soaked in a Kong™ so that it swells, and doesn't fall out too readily; soaking has the added advantage of a considerably increased moisture content (unsoaked kibble fed long-term may be dehydrating, especially in summer). For ease, pop a kibble-filled Kong™ in a cup or bowl of water, and leave to soak for a few minutes, or, alternatively, feed a quality tinned food or balanced raw meat diet from a Kong.™ Most dogs and puppies learn how to empty a Kong™ very quickly!

On first presentation, pack the Kong™ loosely to encourage your dog or puppy to empty it, tamping down the food a little when he becomes more skilled. Remember

Kongs™ are perfect home-alone tools. Sit the Kong™ in a large mug, fill with kibble, and add water. Leave to soak until the kibble swells, making it harder for your dog to empty the toy. Add pieces of extra tasty meat or cheese for variety.

that a Kong™ does not have to be filled to the top: pushing food down to the far end makes reaching it more challenging. You can also thread a rope through the smaller hole of a Kong,™ knot it, and tie the other end around the bars of a crate, which makes it harder to empty for those dogs who are expert Kong™ users! If accessing the food is too difficult, your dog may give up, so always teach the skills he needs first. Other tools and aids include activity balls and food-dispensing toys that engage the brain in puzzle-solving exercises.

Toys

Dogs love novelty, too! Keep a special toy for your dog that is reserved for when you leave the house. Make sure it is appropriate to your dog's breed and temperament: for instance, gun dogs love big, soft toys with large bite areas; your Border Collie or terrier may prefer something that squeaks!

It is not safe to leave certain dogs with toys, however, as some will deglove or chew them. Assess *your* dog's suitability before leaving him with a toy.

When you come home, put away the special, home-alone toy, and swap it for a treat so that your dog does not feel deprived. Perhaps have several special toys to ring the changes and keep him interested.

Should we feed our dog every time we leave him?

If we feel that our dog or puppy is emotionally secure when left alone, it may not be necessary or appropriate to feed him every time we leave him. However, during the initial training period of teaching him to enjoy his time alone, it is good practice to leave him with something every time to distract him.

Once he is comfortable with being left, we may be able to skip feeding him if we are going out only briefly: on these occasions, show him your empty hands to tell him there is no food, or, rather than nothing at all, you may prefer to leave him with his special home-alone toy. Base your decision on how relaxed your dog appears to be on your departure and return. Use the Comfort evaluation worksheet on page 57 to measure his level of emotional comfort. Always err on the side of caution: if unsure, leave your dog with his food-dispensing toy.

continued page 83

Gun dogs derive pleasure simply from holding toys in their mouth.

Some dogs enjoy suckling on toys for comfort.

This dog has been left with a selection of toys, and has made his own comforting 'pillow.'

Some tugs can be threaded through crate roof bars for small dogs and puppies to enjoy playing with inside their crate ...

... though take care to check that the crate is securely placed, and that the tug isn't tied in such a position that your dog can pull over or tip the crate.

Make sure that all toys are strongly made, and specifically for dogs, such as this plaited fleece tug.

This sheepskin tug toy is both soft and durable.

A complete home-alone toolkit: this dog is enjoying a Kong,™ while wearing a bandana sprayed with a relaxing, herbal scent. Lying on a comfortable bed with a favourite toy, soft background music calms his nervous system.

Visual stimulation is reduced by partially closing the blinds, and his focus is entirely on his Kong.™ His face and posture are relaxed and at ease.

Signs that our best friend is happy
A well-balanced dog or puppy will be relaxed about our leaving the house. He would prefer to come with us, of course, but will have the skills to cope without us. Putting on a jacket or shoes that we usually wear when going somewhere without our dog (to work, say), and picking up the car keys will become his signal to rush to the kitchen for his treat or dinner. On our departure he may look up to watch us go, but should be more interested in his treat, toy, or dinner than in following us to the door.

On our return, we should find that our dog or puppy has eaten the food set down for him. He should be pleased to see us, thought not frantically jumping up with the nervous energy often seen in worried dogs, who may also scratch at our legs to gain our attention. Once the excitement of our return has subsided, he will be able to settle down relatively quickly, and will feel secure in his home environment, whether or not we are with him.

A new puppy can be taught to enjoy his own company fairly easily because there's no history of home-alone emotional discontent, such as separation anxiety, to erase. With a more mature dog who has a long history of disappointment at being left, and especially with rescue dogs (who can often be prone to separation anxiety), the same techniques can work, but it may take you longer to modify his behaviour. Please don't put a timescale on your dog's learning, however: it will take as long as it takes.

Throughout this book it is emphasised that prevention is better than cure, and taking action, even in the absence of behaviour issues will help prevent future problems. Whether your dog has been with you for many years, is a rescue dog recently rehomed, or just-arrived puppy, you *can* teach him to enjoy his own company.

Your goal is to build a healthy, balanced relationship, whereby you enjoy each other's company immensely, but are both able to enjoy some time on your own.

Behaviour problems can surface at any time, and illness or old age, accompanied with loss of faculties, can make some dogs more anxious when left. Teach solid *Home alone – and happy!* foundations, and don't become complacent because your dog seems fine, or feel that you don't need the tools any more because he is comfortable. You have created a level of expectation in your dog that you must fulfil. Keep doing what works!

The skills outlined in this book are a proactive training protocol for every dog owner and every dog: don't wait until a problem arises before teaching them. Whatever our dog's age it is never too late to enrich his home environment, and teach him how to enjoy his own company; thus giving him the gift of a good quality of life, even when we are apart.

A dog who is able to enjoy his own company when alone will be emotionally well-balanced and content.

6: Assessment and worksheets

Comfort evaluation worksheets

Observing and assessing your dog's emotional state and his behaviour will help to determine his level of comfort, and give some idea of how long it may take him to learn the new skills he needs.

Use Comfort evaluation worksheet 1 to initially assess how your dog is coping currently when you leave the house. He may display none, some, most, or all of the behaviours listed, and your observations will act as a benchmark.

As your dog improves, you should see fewer behaviours from this worksheet, with a corresponding increase in 'no' answers, as your dog's home-alone skills increase.

Answering 'yes' to most or all of the questions in Comfort evaluation worksheet 2 is your ultimate goal, and, once you have implemented home-alone training, you should see an increase in the number of 'yes' answers.

Choose the same day of the week for each assessment, and record your findings in the tables. Answer the questions in both worksheets in each session, and compare your findings over a four-week period. (We suggest you photocopy the worksheets here, and the forms on page 86.)

REMEMBER!
Your dog does NOT have to show any of these behaviours to benefit from the behaviour modification advice in this book, which can be used to improve his quality of life and well-being, even in the absence of any anxiety. If your dog is, however, showing one or more of these behaviours, he may be worried about being left alone.

Begin your home-alone training as soon as possible, and assess your dog once a week on the same day of the week, for at least a four-week period; longer, if necessary. Give him time to re-learn/learn that being alone can be pleasurable.

Comfort evaluation worksheet 1 – How comfortable is my dog? Week ....

- If my dog is dozing and I leave the room, he gets up and follows me   ❑yes   ❑no   ❑sometimes

- My dog follows me to the bathroom   ❑yes   ❑no   ❑sometimes

- My dog follows me upstairs   ❑yes   ❑no   ❑sometimes

- My dog cries or barks if a door is closed between us   ❑yes   ❑no   ❑sometimes

- My dog looks worried and disappointed when I leave the house, and maybe runs to the window to watch me go   ❑yes   ❑no   ❑sometimes

- My dog cries or barks if he is with another family member, but apart from me   ❑yes   ❑no   ❑sometimes

- If I leave food down when I am out, my dog will not eat it until I return   ❑yes   ❑no   ❑sometimes

- When I return from an outing my dog is frantic with pleasure to see me   ❑yes   ❑no   ❑sometimes

- When I return from an outing my dog follows me around the house   ❑yes   ❑no   ❑sometimes

- When I return from an outing my dog looks anxious and is panting   ❑yes   ❑no   ❑sometimes

Comfort evaluation worksheet 2 – How comfortable is my dog? Week ....

- My dog can settle on his bed in anticipation of a reward   ❑yes   ❑no   ❑sometimes

- My dog can settle when I leave the room   ❑yes   ❑no   ❑sometimes

- My dog can settle when I go upstairs   ❑yes   ❑no   ❑sometimes

- My dog anticipates he will receive something pleasant when I get ready to leave the house   ❑yes   ❑no   ❑sometimes

- My dog looks relaxed when I leave the house   ❑yes   ❑no   ❑sometimes

- My dog is distracted with his home-alone aids when I leave the house   ❑yes   ❑no   ❑sometimes

- My dog is happy to eat his food when I am out of the house   ❑yes   ❑no   ❑sometimes

- My dog is quiet and calm on my return to the house   ❑yes   ❑no   ❑sometimes

- My dog greets me in a friendly yet calm manner on my return   ❑yes   ❑no   ❑sometimes

- My dog is inquisitive about me and what I have with me on my return, but can re-settle quickly   ❑yes   ❑no   ❑sometimes

Home alone – and happy!

Comfort evaluation worksheet 1			
Period covered:			
Add up and note the number of times you answered yes, no or sometimes			
Week 1	Yes	No	Sometimes ....
Week 2	Yes	No	Sometimes ....
Week 3	Yes	No	Sometimes ....
Week 4	Yes	No	Sometimes ....

Comfort evaluation worksheet 2			
Period covered:			
Add up and note the number of times you answered yes, no or sometimes			
Week 1	Yes	No	Sometimes ....
Week 2	Yes	No	Sometimes ....
Week 3	Yes	No	Sometimes ....
Week 4	Yes	No	Sometimes ....

Notes

Visit Hubble and Hattie on the web: www.hubbleandhattie.com
www.hubbleandhattie.blogspot.co.uk
• Details of all books • Special offers • Newsletter • New book news

86

Appendix

Resources and further reading; references

Resources and further reading

The Backwards Brain Bicycle. Destin Sandlin. Smarter Every Day (2015). Video available from: https://www.youtube.com/watch?v=ISYd_ASh3jw. Website: www.smartereveryday.com

Barks & Bunnies. Website: www.barksandbunnies.co.uk. Tel: +44 (0) 800 772 0152. Email: amy@barksandbunnies.co.uk

Honey's Real Dog Food. Website: www.honeysrealdogfood.com. Tel: +44 (0)1672 620260. Email: info@honeysrealdogfood.com

Institute of Modern Dog Trainers. Website: www.imdt.uk.com. Tel: +44 (0)1707 263836. Email: info@imdt.uk.com

International Canine Behaviourists. Website: www.icb.global. Tel: +44 (0)1889 270445 or 07415 885168. Email: info@icb.global

In the Dog House. Website: www.inthedoghousedtc.com. Tel: +34 952 110 243 Email: infospain@inthedoghousedtc.com

Music for Pets. Website: www.theanimalhealer.com. Email: margrit@theanimalhealer.com

Nose2Tail Dog Food. Website: www.nose-2-tail.co.uk. Tel: +44 (0)800 9788 648. Email: info@nose-2-tail. co.uk

Pet Remedy. Website: www.petremedy.co.uk. Tel: +44 (0) 1803 612 772. Email: martyn@petremedy.co.uk

Sachiko Eubanks Photography. Website: www.sachikoeubanksphoto.com. Email: sachikoeubanksphoto@gmail.com

Snuffle Mats™. Website: www.snufflemats.co.uk. Tel 07469 146723. Email: info@snufflemats.co.uk

Through a Dog's Ear. Website: www.throughadogsear.com. Tel: (800) 788 0949 (USA only) or (541) 482 2134. Email: customersuppport@throughadogsear.com

Tug-E-Nuff Dog Gear. Website: www.tug-e-nuff.co.uk. Tel: +44 (0)1395 488088. Email: contactus@tug-e-nuff.co.uk

Through a Dog's Ear: Using Sound to Improve the Health and Behavior of Your Canine Companion by Joshua Leeds and Susan Wagner, DVM, MS. Publisher: Sounds True, 2008. This book includes a starter CD with excerpts from Calm Your Canine Companion, Vol 1 and Music for the Canine Household

References

Animals in Translation: The Woman who Thinks Like a Cow. Dr Temple Grandin. Bloomsbury Publishing (2006). ISBN 978-0-7475-6669-4

BioAcoustic Research & Development (BARD) Canine Research Summary. Joshua Leeds, Lisa Spector & Susan Wagner (2005-2007). Available from www.throughadogsear.com/research/

Dog Body, Dog Mind: Exploring Canine Consciousness and Total Well-Being. Dr Michael W Fox. The Lyons Press (2007). ISBN 978-1-59921-045-2

Home alone – and happy!

Dogs Detect Human Emotions in our Faces. Roger Abrantes. Podcast (2011). Available from https://www.facebook.com/video/video.php?v=10150123036567425&comments

Dominance in Dogs: Fact or Fiction? Barry Eaton. Dogwise Publishing, Direct Book Service Inc (2010 2nd edition). ISBN 978-1-929242-80-1

Don't Shoot the Dog. Karen Pryor. Omnia Books Ltd (2002). ISBN 1-86054-238-7

Efficacy of Dog Appeasing Pheromone in Reducing Stress and Fear-related Behaviour in Shelter Dogs. Elaine Tod, Donna Brander & Natalie Waran. Elserver (2005). Applied Animal Behaviour Science, Volume 93, issue 3, pages 295-308.

For the Love of a Dog: Understand Emotion in You and Your Best Friend. Patricia McConnell. Ballantine Books (2007). ISBN 978-0-345-47715-6

Give Your Dog a Bone. Dr Ian Billinghurst. SOS Print & Media Group (1993). ISBN 0-646-16028-1

Handbook of Behavior Problems of the Dog and Cat. Dr Gary Landsberg, Dr Lowell Ackerman & Dr Wayne Hunthausen. Elsevier (2003 2nd edition). ISBN 0-7020-2710-3

Honey's Natural Feeding Handbook for Dogs. Jonathan Self. The Mammoth Publishing Company (2015 2nd edition). ISBN 978-0-9570753-0-6

I'll be Home Soon: How to Prevent and Treat Separation Anxiety. Patricia McConnell. McConnell Publishing (2010). ISBN 1-891767-10-0

Oxford Dictionary of Animal Behaviour. David McFarland. Oxford University Press (2006). ISBN 978-0-19-86721-2

Penguin Dictionary of Psychology. Arthur S Reber, Rhianon Allen & Emily S Reber. Penguin Books Ltd (2009 4th edition). ISBN 978-0-141-03024-1

Pet Remedy: How It Works. Martyn Barklett-Judge (2014). Available from http://www.petremedy.co.uk/why-it-work/

Teaching with Reinforcement: for every day and in every way. Kay Laurence. Learning About Dogs Ltd (2009). ISBN 1-904116-335

The Emotional Lives of Animals. Marc Bekoff. New World Library (2007). ISBN 978-1-57731-502-5

The Emotional Brain: The Mysterious Underpinnings of Emotional Life. Joseph Ledoux. Orion Books (2003 2nd edition). ISBN 0-75380-670-3

The Inside of a dog: What Dogs See, Smell and Know. Alexandra Horowitz. Simon & Schuster (2009) ISBN 978-1-84737-347-2

Through a Dog's Ear: Using Sound to Improve the Health and Behaviour of your Canine Companion. Joshua Leeds & Susan Wagner. Sounds True (2008). ISBN 978-1-59179-811-8

Woody

Phoebe

Tess

Tupacic

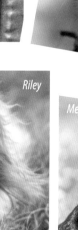
Skye

Tilly

Riley

Meg

Ruby

Lara

Mac

Oscar

Nellie

Malt

Merlin

Martha

Tyler

George

Digby

Hugo

Darcy

Tilly

Aston

Cedar

Button

Betty

Daniel

Cotton

Barnaby

International Canine Behaviourists

ICB

Welfare never compromised

Supporting owners, professionals and dogs

International organisation

Through our world-wide membership, we offer advice to dog owners to help overcome unwanted behaviours, provide continuing education for professionals and support members in their career choice.

Canine specialists

Our members have 'hands on' experience, theoretical knowledge and academic qualifications. We pledge never to use any approach that may harm a dog's physical and mental wellbeing.

Behaviourists' continuing education

As a new graduate you may know the theory, but you don't have the experience! We are committed to supporting your journey to full membership... ...and beyond.

Welfare never compromised !

www.icb.global | info@icb.global

SnuffleMats™ are the perfect way to stimulate your dog's foraging instincts, engaging mind and body, and encouraging natural skills

Why your dog needs SnuffleMats™

- Dogs love to forage for food and treats!
- Mentally stimulating and boredom-busting
- Ideal for dogs on restricted exercise

66 SnuffleMats bring all the benefits of foraging for food into the home. I highly recommend them as an environmental enrichment tool for dogs, and when introducing a puppy to grooming to build up a positive reinforcement history of being touched 99

Kate Mallatratt,
Canine Behaviourist, www.icb.org

www.snufflemats.co.uk

info@snufflemats.co.uk | 07469 146723

Index

Adaptil™ 69
Adrenaline 69
Anticipation 38
Arousal 63
Assessment 84

Bandana 70
Beds 59, 60
Behaviour 14
Bones (raw) 50, 74

Clothing 70
Crate 26, 50, 64, 74

Daily routine 38
Departure 32, 38
Dominance 21
'Down' 52-55

Eating 43
Emotions 22, 23-32
Endorphins 23
Environment 59, 74
Essential oils 70

Family 43
Feeding 40, 43-52, 75
Food ingredients 48

Food-dispensing toys 47-49, 74
Foraging 47
Fussy eaters 50

Genetics 14
Grandin, Dr Temple 70
Guarding 51, 52

Healthy dogs 13, 83
Hearing 25, 69

Independence 40-58

Kong™ 7, 8, 74, 75

Mealtimes 40
Mirroring 20
Movement restriction 64
Multi-dog households 16, 33
Muscle memory 13, 14
Music 66, 68, 69, 74

Neurotransmitter 69

Older dogs 16, 21
Over-attachment 12

Pack leader 21
Panic attacks/PTSD 9
Pet Remedy™ 69
Play 23, 27
Pressure 70
Problem prevention 13
Puppies 7, 32, 40, 83

Rescue dogs 8
Room temperature 62, 74

Scent 69, 74
Separation anxiety 11-21
'Settle' 52
Sleep 59, 60
Smell 25
SnuffleMats™ 44-46
Stress 26
Stressors 14

Theory of mind 25
Toolbox 72-83
Toys 35, 74, 75

Visual stimulation 74